# YOUR KNOWLEDGE HAS VALUE

- We will publish your bachelor's and master's thesis, essays and papers

- Your own eBook and book - sold worldwide in all relevant shops

- Earn money with each sale

Upload your text at www.GRIN.com and publish for free

**Bibliographic information published by the German National Library:**

The German National Library lists this publication in the National Bibliography; detailed bibliographic data are available on the Internet at http://dnb.dnb.de .



**Imprint:**

Copyright © 2015 GRIN Verlag, Open Publishing GmbH
Print and binding: Books on Demand GmbH, Norderstedt Germany
ISBN: 978-3-668-15271-7

**This book at GRIN:**

http://www.grin.com/en/e-book/315321/an-improved-three-factor-remote-user-authentication-scheme-using-smart

**Manish Shingala, Chintan Patel, Nishant Doshi**

# An Improved Three Factor Remote User Authentication Scheme Using Smart Card. A Review

GRIN Publishing

# AN IMPROVED THREE FACTORE REMOTE USER AUTHENTICATION SCHEME USING SMART CARD: A REVIEW

Manish Shingala, Chintan Patel, Nishant Doshi

# An Improved Three Factor Remote User Authentication Scheme Using Smart Card: A Review


Submitted By

**MANISHKUMAR BHANUBHAI SHINGALA**


Under the Guidance of
**CHINTAN PATEL and NISHANT DOSHI**
Department of Computer Engineering
Marwadi Education Foundation Group of Institutions


A Dissertation Phase-1* Submitted to

Computer Engineering Department
Marwadi Education Foundation Group of Institutions
Affiliated to Gujarat Technological University

For the Partial Fulfillment of Degree of Master of Engineering
In Computer Engineering
December – 2015

Faculty of P.G. Studies & Res. in Engineering & Technology
At & PO : Gauridad, Rajkot-Morbi Road
Rajkot 360 003. Gujarat. India.

*Thesis is modified and updated based on Dissertation Phase-1 Presentation and comments.

# Acknowledgements

I am grateful to numerous local and global "peers" who have contributed towards shaping this Dissertation Phase 1 report.

First and foremost I would like to express my sincere thanks to *Dr. Nishant Doshi* who created my interest in *cryptography*. He was always there to guide, motivate and support me whenever I was stuck. He constantly reminded me to achieve my goal. His observations and comments helped me to establish the overall direction of the research and to move forward with investigation in depth. Irrespective of his busy schedule, he always gave time to listen to my doubts patiently and gave valuable suggestions like a parent. His doors were always open to discuss my doubts anytime.

I owe my deep sense of gratitude to Dissertation Phase 1 report examiners *Dr. Sarang Pande, Dr. Nitul Dutta* for their valuable suggestions and critical comments during presentation of credit as well as progress seminars and also sharing their knowledge which influenced me more to carry out this research work. I am thankful to *Prof. Yogesh Ramani,* for helping us to solve typos and grammatical error throughout the book.

I thank to all my student colleagues for providing fun filled and very informative environment to learn and grow. It is their love and encouragement, which helped me a lot during my research work. I had many memorable moments with them inside and out-side my work. I am grateful to all my *colleagues, M.Tech students, Teaching Assistants* and many others, for being with me in my difficult times and for all the emotional support, care, and fun they provided and who have been kind enough to advise and help in their respective roles. I owe a dept of gratitude to all my friends for their guidance and support. *Mayur Oza and Kishan Makadia* for their thoughtful discussion related to research work that helped me to complete this work in timely fashion. *Harshit Champaneri, Dhara Patoliya, Ruchita Kaneria and Jinita Tamboli* for constant motivation to carry out this research work.

I wish to thank staff of *Department of Computer Engineering, MEFGI* for providing me resources throughout my stay in the college.

Last and most important, I thank my family members. Without their constant support, motivation and love, I would not have been reached so far. I dedicate this research work to my family.

**Manish Shingala**

# Index

# <u>List of Figure</u>

# List of Tables

# Abstract

In today's insecure world that is surrounded by public channels which are insecure for data transfer to each other. So it requires some type of remote user authentication mechanism to verify the legitimate user. In remote user authentication schemes, server checks credential of the user and decides that user is legitimate and genuine or not. Remote user also validates the server for mutual authentication purpose. In today's web -enabled world, two- parameters are most important which are security and privacy. For that many researchers work on this field and proposed a schemes which is based on different aspects likes password table, biometric and smart card for authenticating the remote user. But these are not enough for perfect remote authentication, so a scheme is required which has stronger and accurate then old existing schemes. To make most reliable scheme, it is important to calculate computational and communication cost. Also it requires resistance against many other threats. For these mutual authentication and communication privacy are most essential requirements for remote user authentication scheme. Here, in this work we calculated computational and communication cost of different papers, which would be helpful for researchers for their research.

**Keywords:** *remote user authentication, smart card, biometric, attacks, performance analysis, three factor user authentication*

# 1 Introduction[1]

In today's internet enables computer world, remotely data transfer is a main issue in insecure channel. For this, firstly it is important to authenticate the remotely located user and then check the access rights of that particular user. For this, grooming technology is the Remote User Authentication (RUA) scheme. In current world, many organization or applications in daily usually need a user authentication, such as military, banking, insurance, e-commerce, government etc. Use of Remote Password Authentication to check the authority of the remote user over insecure channel.

Basically authentication process can be carried out via *verification processor* and *identification processor*. In *verification processor* scheme, with stored database, system can check one by one proof of information. But in *identification processor* scheme, system can check one to many proof of information with store database. Traditional system for Remote User Authentication is based on password table. It uses user ID and password or user ID and PIN no. for authentication, but both of have several limitation in their area. In tradition scheme, remote user enter his/her ID and password, these ID and password was send to server on insecure channel and server will check the ID and password to compare with password table. If it exists in password table, then server authenticate the remote user and it will send the authentication message to remote user via insecure channel. At that point new in this scheme, smart card is present. All the authentication information stored in smart card and then authenticate the remote user via smart card. But there are mainly two obstacles in tradition systems are (1) Server maintains the password table, so administrator of server will know the ID and password of all remote user, (2) An intruder can impersonate a legal user via stealing the ID and password from password table. And these techniques have become weak and are sensitive to various kind of attacks.

Authentication system has mainly two type of mechanism are (1) two - factor authentication system, (2) three -factor authentication system. In two – factor user authentication system, user need only two components, like smart card and password. But, in three - factor authentications system, user needs three components to authenticate him/her. $1^{st}$ is known like password, $2^{nd}$ is processed like smart card and $3^{rd}$ is one like biometric.

A smart card is an equipment which has embedded circuit which store the authentication information of the user. Biometric scheme has different merits like to more security, uniqueness, permanent, etc. Biometric scheme has different demerits like to not 100 % perfect matching, not modifiable, loss of body component [28]. Due to its pluses, Biometric based schemes are finding its key role in Remote User Authentication scheme and it includes several advantages against traditional password table based scheme. Based on these, many researchers can proposed their papers in the field of remote user authentication scheme using smart card [2] – [18].

Existing scheme has many security downfall and they are sensitive to several kind of attacks, so there are fail to serve all need of the ideal RUA scheme. The perfect ideal RUA scheme which acquires all the security focal points of the current scheme. Ideal RUA scheme accomplish and resist every single conceivable objective and attacks. Ideal RUA should not store any type of password or password table in server. Authentication information is not transmitted as plaintext in insure channel. Also the user has free rights to change the authentication password easily. If any remote user enters wrong user ID or password or biometric then server should evidently find that and informs the user. More of this, one session key was also established between the user and server. If session key was steal, and attacker can use that session key for login then server easily identified that it is an attacker and it was not grant to access the rights. The ID of the user is dynamically change due to login request

---

[1] The content of this book is based on the suggestions received in Dissertation Phase I presentation and report towards master degree of Mr. Manish Shingala.

is transmit on public channel. The perfect scheme much has low computational cost and communication cost for efficient and practical use. Also it resists the all possible attacks and also support multi-server network environment.

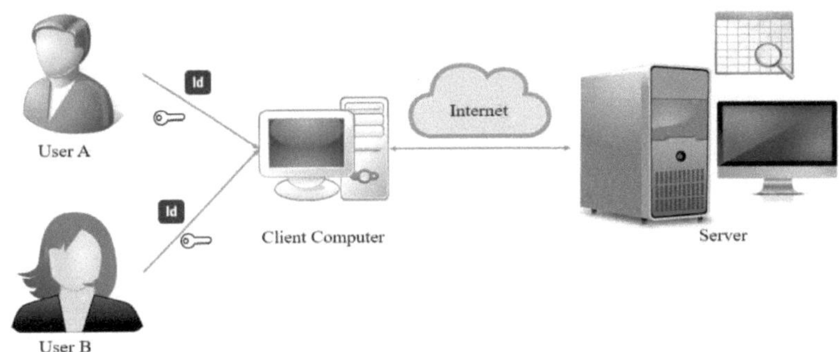

**Fig. 1.1 Traditional scenario of Remote User Authentication Using Password Table**

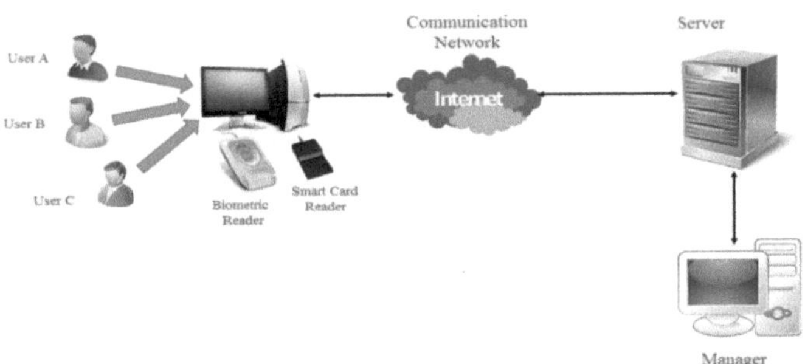

**Fig. 1.2 Remote User Authentication Using Smart Card**

In traditional scenario, the user enter ID and password and he/she can authenticated remotely as shown fig 1.1. And in remote user authentication using smart card, user 1$^{st}$ enter ID, then password and then enter smart card to client side machine and all credentials are sent to remote server for authentication as shown fig 1.2.

Based on the existing approach the Remote User Authentication scheme can be categorized as the password-based, the public-key encryption, the ID-based, the symmetric encryption and the hybrid. In this book, a work is done depend on the prevailing schemes i.e. RSA-based, ELGAMAL-BASED SCHEME and Hash-based.

The documentation utilized as a part of this paper are notice in beneath. Table 1.1.1 present the number of attacks on which this survey is done. There are many types of attack listed by Trupil et al. [20]. Table 1.1.2 gives the different types of function in the performance analysis tables.

**Table 1.1 List of Attacks**

| NOTATIONS | DESCRIPTION |
| --- | --- |
| A1 | Replay Attack |
| A2 | Masquerading or Impersonation Attack |
| A3 | Parallel Session Attack |
| A4 | Mutual Authentication Problem |
| A5 | Man-in-Middle |
| A6 | Forgery Attack |
| A7 | Stolen Session Key |
| A8 | Insider Attack |
| A9 | Password Guessing Attack |
| A10 | User Anonymity |
| A11 | Stolen Smart Card |
| A12 | Session Key Agreement |
| A13 | Server Spoofing Attack |
| A14 | Denial of Services |
| A15 | Information Forward Secrecy |
| A16 | Known Session Specific Temporary information Attack |

**Table 1.2 Different types of Functions for Performance Analysis**

| FUNCTION | DESCRIPTION |
| --- | --- |
| $H(\cdot)$ | Hashing of Biometric |
| $h(\cdot)$ | One-Way Hash Function |
| $\oplus$ | XOR |
| $\parallel$ | Concat Operation |
| e | Exponential Operation |

## 1.1  Our Contribution

In this work we reviewed the state-of-art in remote authentication schemes using smart card. For this, we reviewed the recent schemes Truong et al. [9], Wen et al. [13], Sarvabhatla et al. [14] and Chaturvedi et al. [18] to show the study in this work.

## 1.2  Organization of the book

In the second section, we have discussed literature survey of different remote user authentication schemes with its preliminary and different kind of attacks. In the third section, we mentioned the code for time for each operation. In the fourth section, we have made comparison of computational and communication cost of each paper and in section five, we also discuss different types of attacks on communication channel.

# 2 Literature Survey

In 1981, Leslie Lamport [1] first time suggest "Password Authentication with Insecure Communication". It is Remote User Authentication Scheme based on one-way function to encrypt password in public communication and it is implemented in microcomputer in the user's terminal. This scheme uses password table to store password of the remote user. So it was vulnerable to many kind of attacks. To solve many issues of Lamport's scheme, Hwang et al. proposed authentication scheme in 1990 [21], the benefit of their scheme is that it can't store the password table and verifier table at server side, so the issue of stolen password table at server side is resolve.

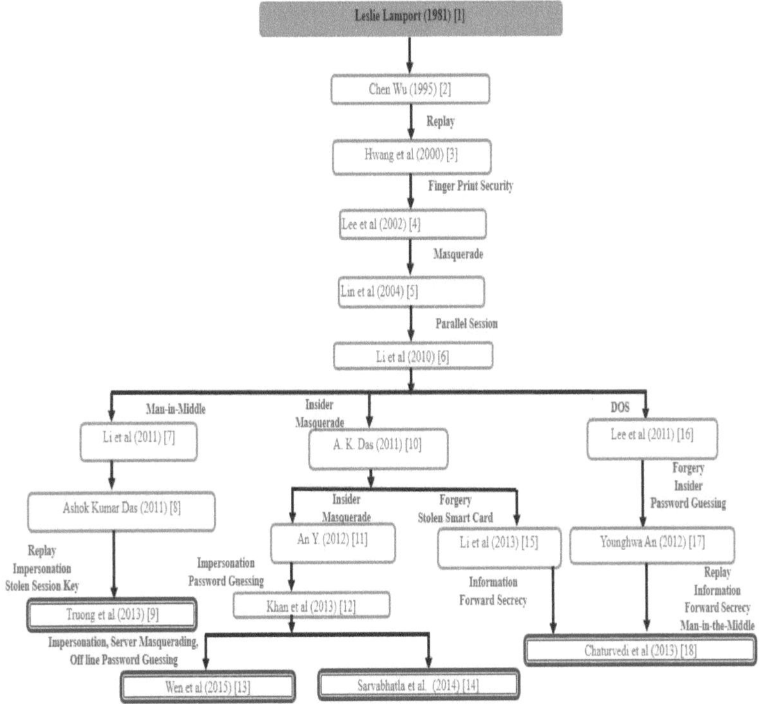


**Fig. 2.1 Literature Survey**

Now shown in Fig. 2.1, in 1995, Chen et al. proposed "Remote login authentication scheme based on a geometric approach" [2], which is based on simple geometric properties on the Euclidean plane. And it has some weakness in the security. In their scheme, it eliminates the use of password table and they use smart card to store authentication information and they give facility to change login password easily to the remote user.

## 2.1 ELGAMAL-BASED SCHEME

In 2000, Hwang et al. [3] found the weakness of Chen et al.'s [2] scheme and it introduced the new Remote User Authentication scheme based on ElGamal public key cryptosystem [22-24]. Hwang et al. [3] claimed that their scheme was secure against replay attack and they perform the Remote User Authentication without using of password table. Their scheme also needs just one private key for communication. Their scheme was vulnerable to different kind of attack many researchers find that attack like DOS, Insider attack, Masquerading attack, Man-in-Middle attack etc.

Later in 2002, Lee et al. recommended "Fingerprint-Based Remote User Authentication Scheme using Smart Card" [4]. Lee et al.'s scheme was based on minutia extraction and matching [25]. When the remote user can input fingerprint that time different map of minutia is made and based on this they generate one-time random number for ElGamal public key cryptosystem using map. With the use of biometric they make their scheme more reliable. In their scheme, they use two secret for authentication, but the drawback in their scheme is that if one secret key out of two was disclosed then the system was vulnerable to attack and never again be kept secure i.e. the legitimate user can act like another legal user easily. So, Lee at al.'s scheme was not secure scheme.

In 2004, Lin et al. use the concept of biometric of Lee et al.'s [5] scheme and recommended "A flexible biometrics remote user authentication scheme". They discovered that Lee et al.'s [5] scheme was vulnerable to impersonation attack and propose new remote user authentication scheme based on ElGamal public key cryptosystem and they use fingerprint verification for authenticate the remote user. And they provide the facility of to change and choose password conveniently and useful for high security require application. Their scheme only maintains one secrete key rather two compared to Lee et al.'s scheme without password table.

**Table 2.1 Performance analysis of ElGamal Based Schemes**

| Papers | | $H(\cdot)$ | $h(\cdot)$ | $\oplus$ | $\parallel$ | E |
|---|---|---|---|---|---|---|
| Hwang et | R | - | - | - | - | 1 |
| al. | L | - | 1 | 1 | - | 3 |
| (2000) | A | - | 1 | 1 | - | 2 |
| [3] | P | - | - | - | - | - |
| Lee et al. | R | - | - | - | - | 2 |
| (2002) | L | - | 1 | 1 | - | 3 |
| [4] | A | - | 1 | 1 | - | 2 |
| | P | - | - | - | - | - |
| Lin et al. | R | - | 1 | 2 | - | 1 |
| (2004) | L | - | 2 | 3 | - | 2 |
| [5] | A | - | 1 | 1 | - | 2 |
| | P | - | 2 | 4 | - | 1 |

**Table 2.2 Requirement of survey comparison between ElGamal Based Scheme**

| Ref. No.<br>Attacks | 2 | 3 | 4 | 5 |
|---|---|---|---|---|
| A1 | Y | | | |
| A2 | | | Y | |
| A3 | | | | Y |
| A4 | | | | |
| A5 | | | | |
| A6 | | Y | | |
| A7 | | | | |
| A8 | | | | |
| A9 | | | | |
| A10 | | | | |
| A11 | | | | |
| A12 | | | | |
| A13 | | | | |
| A14 | | | | |
| A15 | | | | |
| A16 | | | | |

Y = Attack is perform


## 2.2  HASH-BASED SCHEME

In 2010, Li et al. discovered "An efficient biometrics-based remote user authentication scheme using smart cards" [6]. In which the security is depend on biometric verification, one-way hash function and smart card so it is three – factor user authentication scheme. Many RUA uses timestamps to disable the replay attack and for these they need to synchronize clock. But there is no requirement of synchronizing the clock in Li et al.'s scheme because they uses random number instead of timestamps and server does store these random number. And their scheme provides non-repudiation because of employing personal biometric. They identified that Lin et al.'s [5] scheme is unsecured against parallel session attack.

But in 2011, Li et al. found Man-in-Middle attack, A.K. Das found insider and masquerade attack and Lee et al. found denial of service attack on Li et al.'s [6].

First we discussed Li et al.'s scheme. They proposed "Cryptanalysis and improvement of a biometrics-based remote user authentication scheme using smart cards" in 2011 [7]. Their scheme keeps the merit of Li et al.'s [6] scheme and withstand the weakness of old scheme. Addition to old scheme in a new scheme provides session key agreement between the remote user and server after authentication phase was completed. In new Li et al.'s [7] scheme they also assume that R is trusted third party and it will choose two numbers and one number is secret key between server and registration center and second number is secret key between server and remote user.

But new Li et al.'s [7] still have some vulnerability like it does not maintain strong authentication in Login and Authentication phase and user can't update new password which is identified by Ashok Kumar Das and propose new scheme "CRYPTANALYSIS AND FURTHER IMPROVEMENT OF A BIOMETRIC-BASED REMOTE USER AUTHENTICATION SCHEME USING SMART CARDS" in 2011 [8]. This scheme also keeps the merit of Li et al.'s [7] scheme and withstand the weakness of Li et al.'s scheme and A. K. Das's scheme provide strong mutual authentication after successful completion of authentication phase and their scheme always update the new password in the smart card with user friendly environment without the help of registration Centre.

Later in 2013, Truong et al. found replay attack, stolen session key attack and impersonation attack in Ashok Kumar Das's [8] scheme and discover "IMPROVED ID-BASED REMOTE USER AUTHENTICATION SCHEME USING SMART CARD" [9]. They use some extra random value to resist the attacks and decries the computational cost for better performance. Plus on these, they uses three-way challenge-response handshake technique to oppose replay attack. And it is necessary to have equal session key with the user and the server after authentication phase.

As mention above, A. K. Das found weakness insider attack and masquerade attack from Li et al.'s [6] scheme and proposed "Analysis and improvement on an efficient biometric-based remote user authentication scheme using smart cards" in 2011 [10]. Their scheme contain all benefits of Li et al.'s [6] scheme and in order to overcome the weakness they use user's biometric template for verify the user and when the remote user enter his/her biometric at that time their biometric template match with stored biometric pattern and if match then access is granted otherwise not. And their scheme provide mutual authentication, freely password changing, reduce computational cost and without synchronization clock with compared to Li et al.'s scheme.

Later, An Y. found insecurity like impersonation attack, server masquerade attack, password guessing attack, insider attack and weakness of mutual authentication in A. K. Das's [10] scheme and proposed new scheme "Security Analysis and Enhancements of an Effective Biometric-Based Remote User Authentication Scheme Using Smart Cards" in 2012 [11]. They find weakness of Das's scheme by assuming an attacker get secret information of smart card by power consumption [26] [27] and propose new scheme which eliminates the weakness of Das's scheme. An Y.'s scheme have only three phase, it can't mention any type of password change.

But still An Y.'s [11] have weakness like password guessing attack, impersonation attack mutual authentication and user anonymity fined by Khan et al. and propose new scheme "An Improved Biometrics-Based Remote User Authentication Scheme with User Anonymity" in 2013 [12].

Registration phase and Password Change phase carried out in secure channel while Login phase and Authentication phase carried out from insecure channel in An Y.'s scheme. Addition to these, server maintains two type of secret values for improve the security. The importance of that scheme is minor increase of 2 hash functions in computational cost to achieve great performance as compared to previous scheme. Khan et al. consider the password change phase which can't mention in previous An Y.'s scheme and improve performance. Khan et al. claim that his scheme is secure even if legitimate user's secrete information is leaked.

Sarvabhatla et al. found stolen smart card attack, impersonation attack, server spoofing attack and parallel session attack from Khan et al.'s [12] scheme and propose "A Secure Biometrics-Based Remote User Authentication Scheme for Secure Data Exchange" in 2014 [14]. New scheme provide security even if possible secret information of legitimate user was leaked including smart card. Their scheme can't consider the Password change phase and they consider all other phase only.

In 2015, Wen et al. found user anonymity, off-line password guessing attack, server masquerading attack, impersonation attack and weakness in mutual authentication from Khan et al.'s [12] scheme and proposed "Analysis and Improvement on a Biometric-Based Remote User Authentication Scheme Using Smart Cards" [13]. They are use symmetric-key cryptographic techniques (e.g. AES) in order to improve the security.

Li et al. found weakness forgery attack, stolen smart card attack and session key agreement from A. K. Das's scheme and discover "Applying biometrics to design three-factor remote user authentication scheme with key agreement" in 2013 [15]. Their scheme is based on discrete logarithm problem. Their scheme contain benefit of Das's scheme and secure against Das's scheme's attack and provide security. In their scheme the registration centre chooses three parameters to security in which 2 parameters are prime number and $3^{rd}$ is a generator.

As discuss above, Lee et al. found vulnerability and can't achieve session key agreement from Li et al.'s [6] scheme and proposed "Improvement of Li-Hwang's Biometrics-based Remote User Authentication Scheme using Smart Cards" in 2011 [16]. To resist the Denial of Service attack of Li et al.'s scheme Lee et al. assume that the remote user's authenticity is verify in earlier stage means verify user and server each other before authenticate the login request of the user by server.

Later Y. An found password guessing attack, forgery attack, insider attack and weakness in session key agreement from Lee et al.'s [16] scheme and proposed "Improved Biometrics-Based Remote User Authentication Scheme with Session Key Agreement" in 2012 [17]. Their scheme is based on discrete logarithm problem. Y. An claim that their scheme is secure even if the secret information of legitimate user was stolen including smart card. Their scheme contain only three phase namely registration phase, login phase and authentication phase.

But in 2013, Chaturvedi et al. found vulnerability like replay attack, information forward secrecy, user anonymity, session specific temporary information attack from and stolen smart card attack from Li et al.'s [15] scheme and vulnerability like session specific temporary information attack, information forward secrecy, replay attack, man-in-middle attack and user anonymity from Y. An's [17] scheme. Chaturvedi et al. mention weakness of both scheme and proposed new remote user authentication scheme "Improved Biometric-Based Three-factor Remote User Authentication Scheme with Key Agreement Using Smart Card" [18]. Their proposed scheme was eliminate the vulnerably of Y. An's scheme and Li et al.'s scheme and his scheme is based on discrete logarithm problem.

**Table 2.3 Performance analysis of Hash Based Schemes**

| Papers | | $H(\cdot)$ | $h(\cdot)$ | $\oplus$ | $\|$ | E |
|---|---|---|---|---|---|---|
| Li et al. | R | 1 | 3 | 1 | 3 | - |
| (2010) | L | 1 | 2 | 2 | 2 | - |
| [6] | A | - | 5 | 3 | 5 | - |
| | P | 1 | 3 | 2 | 3 | - |
| Li et al. | R | 1 | 3 | 1 | 3 | - |
| (2011) | L | 1 | 4 | 3 | 5 | - |
| [7] | A | - | 7 | 4 | 15 | - |
| | P | 1 | 4 | 2 | 4 | - |
| Ashok | R | 1 | 3 | 1 | 2 | - |
| Kumar Das | L | 1 | 4 | 3 | 5 | - |
| (2011) | A | - | 9 | 6 | 17 | - |
| [8] | P | 1 | 8 | 5 | 8 | - |
| Truong et al. | R | 1 | 5 | 2 | 6 | - |
| (2013) | L | 1 | 4 | 3 | 6 | - |
| [9] | A | - | 7 | 10 | 3 | - |
| | P | 1 | 8 | 5 | 8 | - |
| A K Das | R | 1 | 2 | 2 | 1 | - |
| (2011) | L | 1 | 2 | 3 | - | - |
| [10] | A | - | 8 | 3 | 5 | - |
| | P | 1 | 3 | 4 | 1 | - |
| An Y. | R | 1 | 3 | 6 | 1 | - |
| (2012) | L | 1 | 3 | 5 | 1 | - |
| [11] | A | - | 6 | 3 | 8 | - |
| | P | 1 | - | - | - | - |
| Khan et al. | R | 1 | 3 | 7 | 4 | - |
| (2013) | L | 1 | 2 | 9 | 4 | - |
| [12] | A | - | 7 | 4 | 10 | - |
| | P | 1 | 2 | 8 | 4 | - |

| | | | | | |
|---|---|---|---|---|---|
| Wen et al. | R | - | 4 | 10 | 2 | - |
| (2015) | L | - | 2 | 3 | 3 | - |
| [13] | A | - | 9 | - | 26 | - |
| | P | - | 4 | 7 | 2 | - |
| Sarvabhatla | R | 1 | 11 | 16 | 14 | - |
| et al. | L | 1 | 10 | 11 | 13 | - |
| (2014) | A | - | 12 | 7 | 29 | - |
| [14] | P | - | - | - | - | - |
| Li et al. | R | - | 4 | 1 | 4 | - |
| (2013) | L | - | 3 | 2 | 3 | - |
| [15] | A | - | 8 | 3 | 10 | - |
| | P | - | 2 | - | 1 | - |
| Lee et al. | R | 1 | 3 | 1 | 3 | - |
| (2011) | L | 1 | 3 | 2 | 2 | - |
| [16] | A | - | 8 | 3 | 11 | - |
| | P | 1 | 3 | 2 | 3 | - |
| Y An | R | 1 | 3 | 3 | 2 | - |
| (2012) | L | 1 | 2 | 3 | 2 | 1 |
| [17] | A | - | 1 | 3 | 5 | 1 |
| | P | - | - | - | - | - |
| Chaturvedi | R | - | 2 | 2 | 4 | - |
| et al. | L | - | 3 | 3 | 7 | 3 |
| (2013) | A | - | 4 | 1 | 14 | 4 |
| [18] | P | - | 4 | 3 | 6 | - |

**Table 2.4 Requirement of Security Comparison between Hash Based Schemes**

| Attacks \ Ref. No. | 6 | 7 | 8 | 10 | 11 | 12 | 15 | 16 | 17 |
|---|---|---|---|---|---|---|---|---|---|
| A1 | | | Y | | | | Y | | Y |
| A2 | Y | | | Y | | Y | | | |
| A3 | | | | | | Y | | | |
| A4 | Y | Y | | Y | Y | Y | | | |
| A5 | Y | | | | | | | | Y |
| A6 | | | Y | Y | Y | Y | | Y | |
| A7 | | | Y | | | | | | |
| A8 | Y | | | Y | | | | Y | |
| A9 | | | | Y | Y | Y | | Y | |
| A10 | | | | | Y | Y | Y | | Y |
| A11 | | | | Y | | Y | Y | | |
| A12 | | | | Y | | | | Y | |
| A13 | | | | | | Y | | | |
| A14 | Y | | | | | | | | |
| A15 | | | | | | | Y | | Y |
| A16 | | | | | | | Y | | Y |

Y = Attack is performed

# 3  Code for finding Program Code

In this section, we demonstrate the code for different type of operation. For this, we use windows 7 Ultimate 64-bit operating system which has Intel core i5 $2^{nd}$ generation processor and 4 GB RAM and eclipse tool, and the same operations are perform on Samsung Android Galaxy Grand I9082 smartphone which have Jellybean 4.1.2 operating system, 1 GB RAM and Dual-core 1.2 GHz Cortex-A9 processor and use AIDE tool for get timing of operation.

## 3.1  Programming code for one-way Hash Function (SHA-1)

```java
import java.security.MessageDigest;
import java.security.NoSuchAlgorithmException;
import java.util.*;
import java.util.Random;
class SHA {
      public static void main(String[] args) throws
NoSuchAlgorithmException {
            long endTime=0;
            int randomInt;
            long totalTime;
            Random randomGenerator = new Random();
            for (int idx = 1; idx <= 100; ++idx)
            {
              randomInt = randomGenerator.nextInt(100);
              String string3 = Integer.toString(randomInt);
              long start = System.nanoTime();
              System.out.println(sha1(string3));
              long end = System.nanoTime();
              long end1 = end - start;
              System.out.println(end1);

              endTime = endTime + end1;
            }

            long avg = (endTime/100);
            System.out.println("Total time for 100 iteration is " +
  endTime);
            System.out.println("Average time for SHA operation is " +avg);

      }

      static String sha1(String input) throws NoSuchAlgorithmException {
          MessageDigest mDigest = MessageDigest.getInstance("SHA1");
          byte[] result = mDigest.digest(input.getBytes());
          StringBuffer sb = new StringBuffer();
          for (int i = 0; i < result.length; i++) {
              sb.append(Integer.toString((result[i] & 0xff) + 0x100,
16).substring(1));
          }

          return sb.toString();
      }
}
```

Below is a screenshot for time require to execute the one-way hash function in Fig 3.1 and Fig 3.2

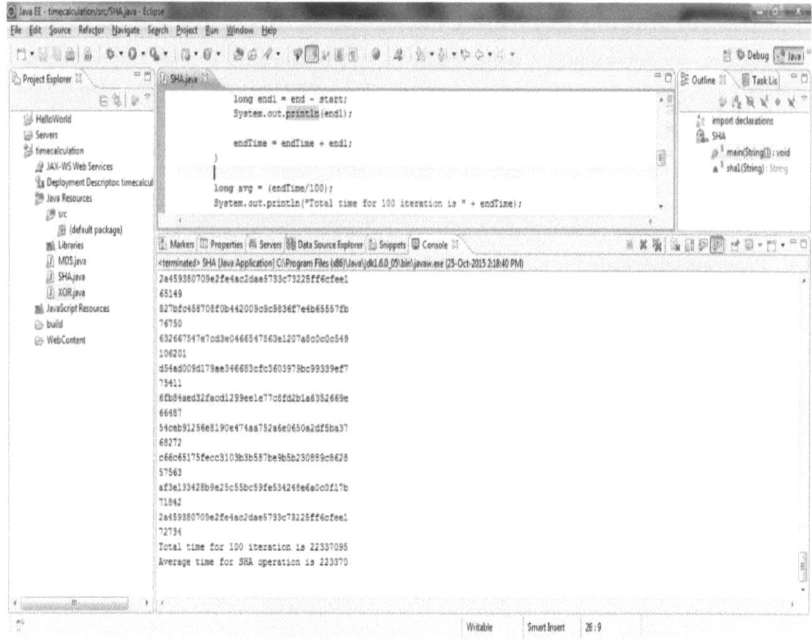

**Fig. 3.1 Timing for One-way hash function in PC**

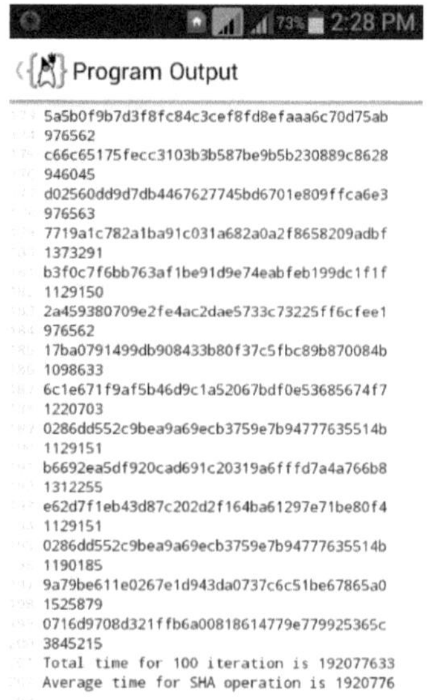


**Fig. 3.2 Timing for One-way hash function in Android Phone**

12

## 3.2  Programming code for XOR operation

```java
import java.util.*;
class XOR {
  public static void main(String args[])
  {

          long endTime=0;
          Random randomGenerator = new Random();
          for (int idx = 1; idx <= 100; ++idx)
          {
            int randomInt = randomGenerator.nextInt(100);
            int randomInt1 = randomGenerator.nextInt(100);

            System.out.println(randomInt);
            System.out.println(randomInt1);
            System.out.println("1st Random:- "
+Integer.toBinaryString(randomInt));
            System.out.println("2nd Random:- "
+Integer.toBinaryString(randomInt1));
            long start = System.nanoTime();
            System.out.println("XOR Result =" +(randomInt ^ randomInt1));
            long end = System.nanoTime();
            long end1 = end - start;
            System.out.println("Timing for "+ idx +" iteration is"
+end1);

            endTime = endTime + end1;

          }
          long avg = (endTime/100);
          System.out.println("Total time for 100 iteration is " +
endTime);
          System.out.println("Average time for XORing operation is "
+avg);
      }
}
```

Below is a screenshot for time require to execute the one-way hash function in Fig 3.3 and Fig 3.4

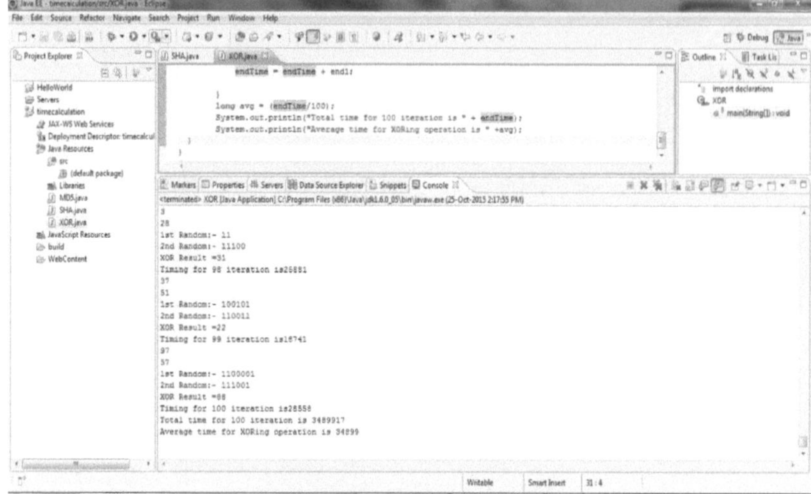

Fig. 3.3 Timing for XOR operation in PC

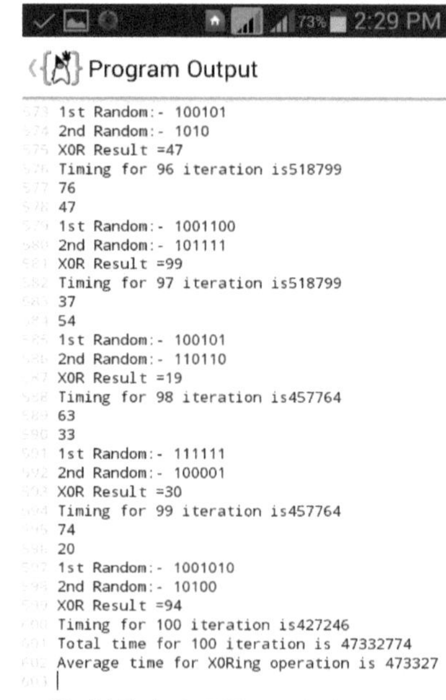


Fig. 3.4 Timing for XOR operation in Android

## 3.3 Programming code for Concat operation

```java
import java.util.*;

public class StringConcat {
        public static void main(String args[]){

                String str1 = "Hello";
                String str2 = " World";

                //1. Using + operator
                long start1 = System.nanoTime();
                String str3 = str1 + str2;
                long end1 = System.nanoTime();
                long total1 = end1 - start1;
                System.out.println("String concat using + operator : " +
str3);
                System.out.println("Timing for iteration is " +total1);
                //2. Using String.concat() method
                long start2 = System.nanoTime();
                String str4 = str1.concat(str2);
                long end2 = System.nanoTime();
                long total2 = end2 - start2;
                System.out.println("String concat using String concat
method : " + str4);
                System.out.println("Timing for iteration is " +total2);
                //3. Using StringBuffer.append method
                long start3 = System.nanoTime();
                String str5 = new
StringBuffer().append(str1).append(str2).toString();
                long end3 = System.nanoTime();
                long total3 = end3 - start3;
                System.out.println("String concat using StringBuffer
append method : " + str5);
                System.out.println("Timing for iteration is " +total3);
        }
}
```

Below is a screenshot for time require to execute the one-way hash function in Fig 3.5 and Fig 3.6

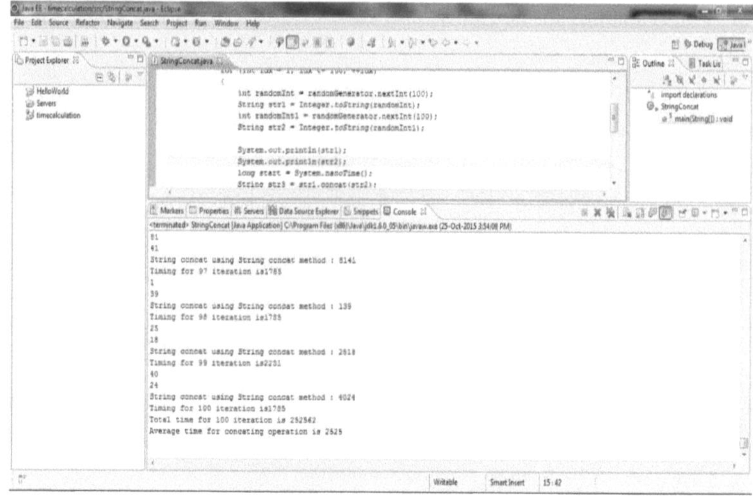

**Fig. 3.5 Timing for Concat operation in PC**

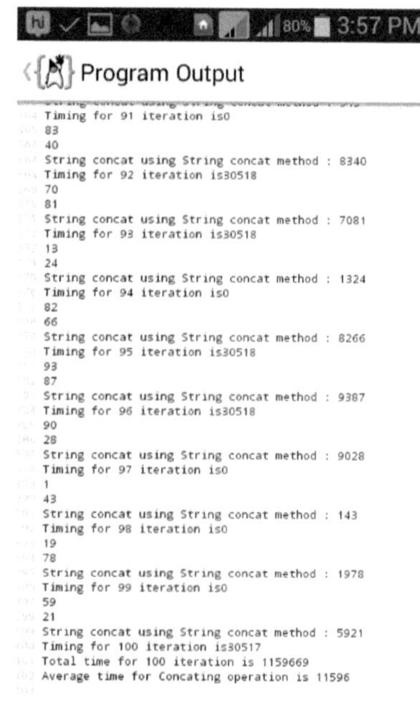


**Fig. 3.6 Timing for Concat operation in Android**

## 3.4  Programming code for Exponential operation

```java
import java.util.*;
class Exponents {
  public static void main(String args[])
   {
                   long endTime=0;
                   Random randomGenerator = new Random();
                   for (int idx = 1; idx <= 100; ++idx)
                   {
                     int randomInt = randomGenerator.nextInt(100);
                     int randomInt1 = randomGenerator.nextInt(100);
                     System.out.println(randomInt);
                     System.out.println(randomInt1);
                     System.out.println("1st Random:- " +Integer.toBinaryString(randomInt));
                     System.out.println("2nd Random:- " +Integer.toBinaryString(randomInt1));
                     long start = System.nanoTime();
                     int z = integerPower(randomInt, randomInt1 );
                     System.out.printf("The exponential form is: %d\n", z);
                     long end = System.nanoTime();
                     long end1 = end - start;
                     System.out.println("Timing for "+ idx +" iteration is" +end1);
                     endTime = endTime + end1;
                   }
                   long avg = (endTime/100);
                   System.out.println("Total time for 100 iteration is " + endTime);
                   System.out.println("Average time for Exponential operation is " +avg);
    }
  public static int integerPower(int randomInt,int randomInt1)
                {
                  int result = randomInt;
                  for (int i = 1; i < randomInt1; ++i)
                  {
                     result = result * randomInt;
                  }
                  return result;
                }
}
```

Below is a screenshot for time require to execute the one-way hash function in Fig 3.7 and Fig 3.8.

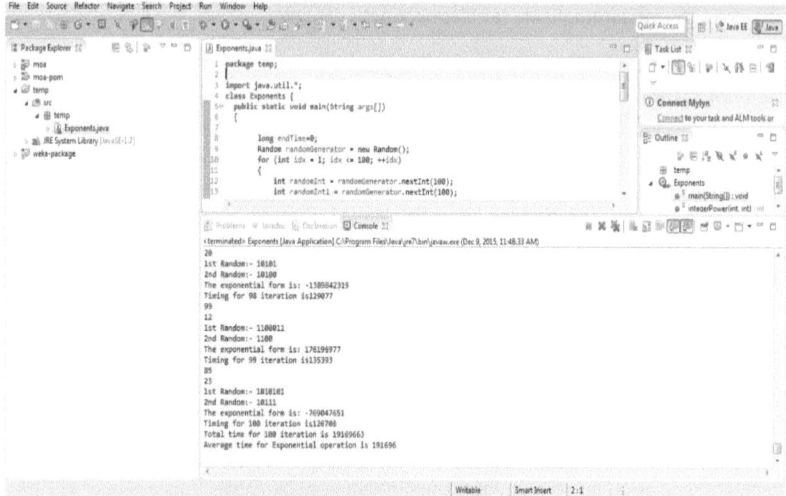

Fig. 3.7 Timing for Exponential operation in PC

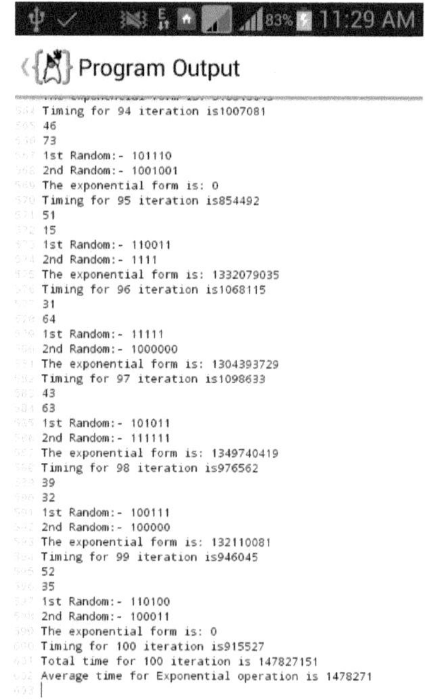


Fig. 3.8 Timing for Exponential operation in Android

# 4  Functionality and Performance Comparison

Here, we compare the functionality of all surveyed paper in this paper tabulated below. For this, we compare the computational cost involved in Registration phase, Login Phase, Authenticate phase and Password Change Phase. For this, we review above Table 2.1 and 2.2 and make performance analysis. Here we consider time complexity for Hash Function, Exclusive OR Function, Concat Function and some scheme contain exponential function. Table 4.2 shows the computational complexity require by ElGamal Based scheme and Table 4.3 shows the computational complexity require by Hash Based scheme. Based on over analysis we measure the time for different type of functions which are tabulated in Table 4.1.

**Table 4.1 Computational cost Analysis of ElGamal Based Schemes**

| Functions | Timing of PC | Timing of Android | Output Bit |
| --- | --- | --- | --- |
| MD5 | 0.20 ms | 1.62 ms | 128 Bit |
| SHA1 | 0.21 ms | 2.20 ms | 160 Bit |
| XOR | 0.033 ms | 0.79 ms | |
| CONCAT | 0.003 ms | 0.01 ms | |
| EXPONENTIAL | 0.1613 ms | 1.56 ms | |

**Table 4.2 Computational cost Analysis of ElGamal Based Schemes**

| Papers | Registration Phase | Login Phase | Authentication Phase | Password Change Phase | Total |
| --- | --- | --- | --- | --- | --- |
| Hwang et al. (2000) [3] | $1T_E$ | $1T_h+1T_x$ $+3T_E$ | $1T_h+1T_x$ $+2T_E$ | - | $2T_h+2T_x$ $+6T_E$ $=\mathbf{0.018}$ |
| Lee et al. (2002) [4] | $2T_E$ | $1T_h+1T_x$ $+3T_E$ | $1T_h+1T_x$ $+2T_E$ | - | $2T_h+2T_x$ $+7T_E$ $=\mathbf{0.021}$ |
| Lin et al. (2004) [5] | $1T_h+2T_x$ $+1T_E$ | $2T_h+3T_x$ $+2T_E$ | $1T_h+1T_x$ $+2T_E$ | $2T_h+4T_x$ $+1T_E$ | $6T_h+10T_x$ $+6T_E$ $=\mathbf{0.0191}$ |

**Table 4.3 Computational cost Analysis of Hash Based Schemes**

| Papers | Registration Phase | Login Phase | Authentication Phase | Password Change Phase | Total |
|---|---|---|---|---|---|
| Li et al. (2010) [6] | $1T_H+3T_h+1T_x+3T_c$ | $1T_H+2T_h+2T_x+2T_c$ | $5T_h+3T_x+5T_c$ | $1T_H+3T_h+2T_x+3T_c$ | $3T_H+13T_h+8T_x+13T_c$ $=\mathbf{0.041}$ |
| Li et al. (2011) [7] | $1T_H+3T_h+1T_x+3T_c$ | $1T_H+4T_h+3T_x+5T_c$ | $7T_h+4T_x+15T_c$ | $1T_H+4T_h+2T_x+4T_c$ | $3T_H+18T_h+10T_x+27T_c$ $=\mathbf{0.083}$ |
| Ashok Kumar Das (2011) [8] | $1T_H+3T_h+1T_x+2T_c$ | $1T_H+4T_h+3T_x+5T_c$ | $9T_h+6T_x+17T_c$ | $1T_H+8T_h+5T_x+8T_c$ | $3T_H+24T_h+15T_x+32T_c$ $=\mathbf{0.098}$ |
| Truong et al. (2013) [9] | $1T_H+5T_h+2T_x+6T_c$ | $1T_H+4T_h+3T_x+6T_c$ | $7T_h+10T_x+3T_c$ | $1T_H+8T_h+5T_x+8T_c$ | $3T_H+24T_h+20T_x+23T_c$ $=\mathbf{0.072}$ |
| A K Das (2011) [10] | $1T_H+2T_h+2T_x+1T_c$ | $1T_H+2T_h+3T_x$ | $8T_h+3T_x+5T_c$ | $1T_H+3T_h+4T_x+1T_c$ | $3T_H+15T_h+12T_x+7T_c$ $=\mathbf{0.023}$ |
| An Y. (2012) [11] | $1T_H+3T_h+6T_x+1T_c$ | $1T_H+3T_h+5T_x+1T_c$ | $6T_h+3T_x+8T_c$ | - | $2T_H+12T_h+14T_x+10T_c$ $=\mathbf{0.032}$ |
| Khan et al. (2013) [12] | $1T_H+3T_h+7T_x+4T_c$ | $1T_H+2T_h+9T_x+4T_c$ | $7T_h+4T_x+10T_c$ | $1T_H+2T_h+8T_x+4T_c$ | $3T_H+14T_h+28T_x+22T_c$ $=\mathbf{0.069}$ |
| Wen et al. (2015) [13] | $4T_h+10T_x+2T_c$ | $2T_h+3T_x+3T_c$ | $9T_h+26T_c$ | $4T_h+7T_x+2T_c$ | $19T_h+20T_x+33T_c$ $=\mathbf{0.102}$ |
| Sarvabhatla et al. (2014) [14] | $1T_H+11T_h+16T_x+14T_c$ | $1T_H+10T_h+11T_x+13T_c$ | $12T_h+7T_x+29T_c$ | - | $2T_H+33T_h+34T_x+56T_c$ $=\mathbf{0.172}$ |
| Li et al. (2013) [15] | $4T_h+1T_x+4T_c$ | $3T_h+2T_x+3T_c$ | $8T_h+3T_x+10T_c$ | $2T_h+1T_c$ | $17T_h+6T_x+18T_c$ $=\mathbf{0.055}$ |
| Lee et al. (2011) [16] | $1T_H+3T_h+1T_x+3T_c$ | $1T_H+3T_h+2T_x+2T_c$ | $8T_h+3T_x+11T_c$ | $1T_H+3T_h+2T_x+3T_c$ | $3T_H+17T_h+8T_x+19T_c$ $=\mathbf{0.058}$ |
| Y An (2012) [17] | $1T_H+3T_h+3T_x+2T_c$ | $1T_H+2T_h+3T_x+2T_c+1T_E$ | $1T_h+3T_x+5T_c+1T_E$ | - | $2T_H+6T_h+9T_x+9T_c+2T_E$ $=\mathbf{0.345}$ |
| Chaturvedi et al. (2013) [18] | $2T_h+2T_x+4T_c$ | $3T_h+3T_x+7T_c+3T_E$ | $4T_h+1T_x+14T_c+4T_E$ | $4T_h+3T_x+6T_c$ | $13T_h+9T_x+31T_c+7T_E$ $=\mathbf{1.206}$ |

$T_H$ = Computational cost for hashing of Biometric

$T_h$ = Computational cost for one-way hash function

$T_x$ = Computational cost for Exclusive OR Function

$T_c$ = Computational cost for Concat Function

$T_E$ = Computational cost for Exponential Function

- Not Applicable

**Note:** Above scheme [15-18] takes extra time of fuzzy generator to take biometric as input and give two string as a output for calculation.

# 5 Different Type of Attacks

This section describes all possible attacks on remote user authentication scheme describe above in Table 1.1.1.

## 5.1 Replay Attack (A1)

When any unauthorized user holds secure information and then retransmits it after some time, then it is called Replay attack. It is a type of delay attack and network type attack and it relate to Man-in-Middle attack. Remote user enters his/her secure ID and password and these credential are send on unsecure channel. At that time any unauthorized attacker will retrieve this credential and replay it after sometimes. To prevent this attack, we use session token and one-time password scheme. Sometimes called Playback Attack.

**Countermeasure:**
To prevent from replay attack, we utilize session token, One Time Password scheme, timestamp [19].

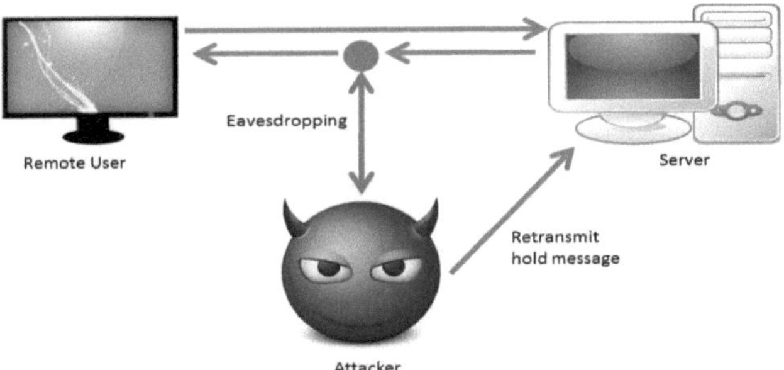


**Fig. 5.1 Replay Attack**

## 5.2  Masquerading Attack (or Impersonation Attack) (A2)

When any attacker claims to be legitimate user of a system to get access to the system or to get higher privileges than they are authorized for, is called masquerade attack. It is a disguise and pursued through the use of pinched logon id and passwords, by finding security gaps in programs, or bypassing the authentication mechanism.

This type of attack basically comes from within an organization or from foreign user via specific connection to the public network. Weak authentication is easy way for masquerade attack and it becomes much easy for an attacker to get access. Once the attacker has identified by server as a legal user, it may have all access to the user's sensitive data.

**Countermeasure:**

To make strong authentication system, improve security for ID and Password be kept secret, we utilize session token, One Time Password scheme, timestamp [19].

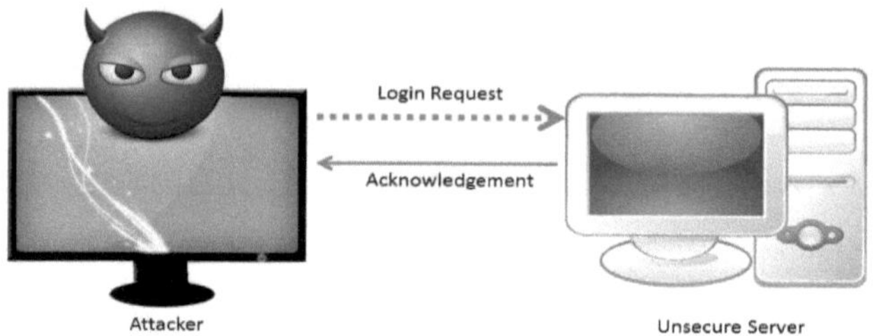


**Fig. 5.2 Masquerading Attack**

## 5.3  Parallel Session Attack (A3)

In parallel session attack a user U, login to server using login request message, after getting successful login server generate session for valid user. Legal adversary E captures a login request from user U and creates a parallel session to masquerade as a user U. It betokens that Parallel Session attack can be applied on parallel processes, methods, tasks.

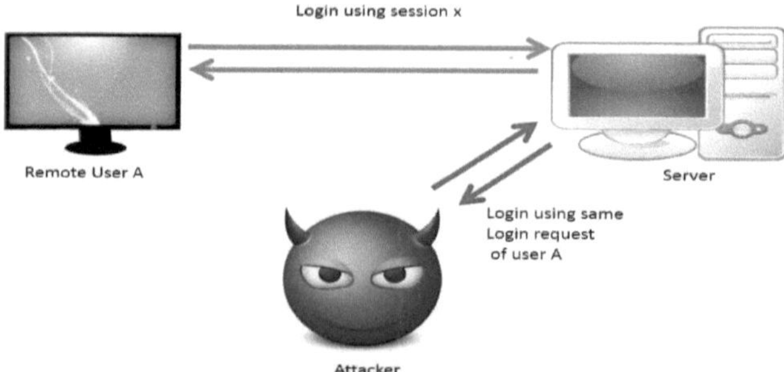


**Fig. 5.3 Parallel Session Attack**

**Countermeasure:**

Use SSL combined with cookie management system. Over hubs prefer switches to avoid this. Cryptography or secure protocol also helps to prevent. The remote connections and incoming connections can be minimized as to solve attacks.

## 5.4  Mutual Authentication (A4)

If in server side or remote user side proper authentication does not perform than many issues will arise, this lack of facility is called Mutual Authentication problem. If any scheme does not provide security against impersonation and masquerading attack, then that scheme will not be able to provide mutual authentication between remote user and server.

**Countermeasure:**

Provide strong authentication mechanism.

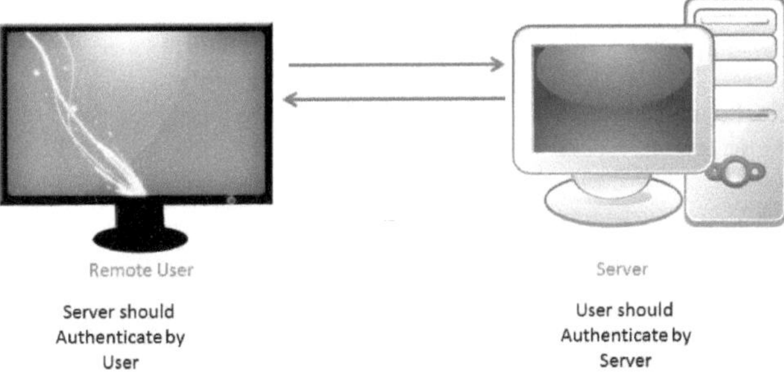


**Fig. 5.4 Mutual Authentication**

## 5.5  Man -In- Middle Attack (A5)

In man-in-the middle attack the attacker secretly intercepts & relays messages of the two parties that believe to be communicating directly with each other. It's form of eavesdropping in which the attacker controls the entire conversation & can also modify each messages.

Sometime it is known as MITM, MitM, MITMA or session hijacking attack. Major security threats to online security is MITM attacks where attacker can capture or manipulate sensitive information during transactions, conversation or transferring data.

**Countermeasure:**

Use SSL for to prevent this attack [19].

**Fig. 5.5 Man –in- Middle**

## 5.6   Forgery Attack (A6)

Valid remote user send login request to server and server should authenticate it and reply acknowledge to the remote user. Now attacker would eavesdrop this processor and make forge of user to server and send login request to server and server successfully authenticate it.

**Countermeasure:**

Use cookie or multi step transaction or synchronization token pattern or client side safe guard or cross side scripting to prevent this attack.

## 5.7   Stolen Session Key by Valid User (A7)

When attacker A is a legitimate user, then A should easily get the session key of another legal user U of the system. Now, attacker A can easily encrypt all messages using U's session key for communication with server after authentication is done by attacker.

**Countermeasure:**

Improve user level security so other user would not see any data of any legitimate user.

## 5.8   Insider Attack (A8)

In the registration phase, if the user's password PW and biometrics Bi are known to the server, the associate of the server may easily get the user's PW and Bi. Thus, the associate of the server act as an attacker can act as the legitimate user to contact the user's other accounts in other server if the user uses the equal password for the other accounts.

**Countermeasure:**

Intrusion Detection System will be helpful to prevent this type of attack. Also designer consider the access control mechanism, logging and monitoring must be strictly maintained [19].

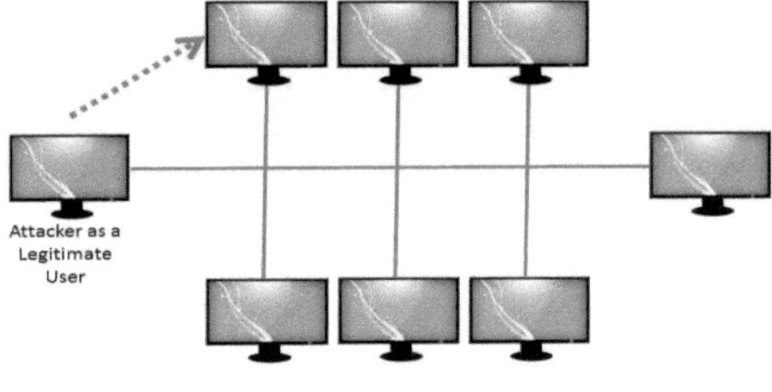


**Fig. 5.6 Insider Attack**

## 5.9   Password Guessing Attack (A9)

Generally, probable passwords are selected through users because probable passwords can be memorized easily by users and users use same passwords for other systems. If Attacker can reveal password of user for one system, then there is possibility to crack another system's password. One fact about password guessing attack is that it does not work always. Password guessing attacks can be categorized into two.

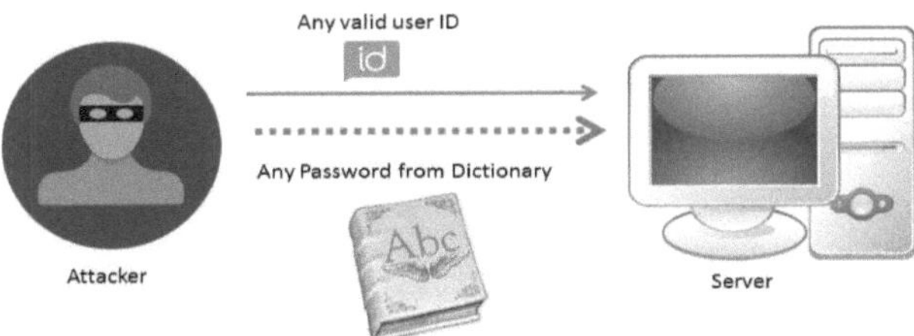


**Fig. 5.7 Password Guessing Attack**

1. Brute Force Attack: A Brute Force attack is a type of password guessing attack in which you have to try each available code, combination, or password until you find the correct one. It requires more time to break the system.

2. Dictionary Attack: A dictionary attack is password guessing attack which utilizes a lexicon of basic words to identify the user's password.

**Countermeasure:**

Avoid by choosing strong password, this strong password must keep alphabets, numbers and special characters. Password should not be in the from of dictionary [19].

## 5.10 User Anonymity (A10)

When user's id is unknown to server, it is called User Anonymity. Main advantage is that remote user will be non-identifiable, untraceable or unreachable. It is a technique to provide privacy or liberty. By using User anonymity the attacker could not identify 'Which users are communicating'. Now, user's identity is associated with login request message, so if this login message is compromised then user's identity is revealed to attacker.

**Countermeasure:**

Secure communication message via some kind of mechanism like encryption techniques.

## 5.11 Stolen Smart Card (A11)

We guess that smart card is lost by user and that smart card is captured by any attacker. Attacker retrieves credential which are placed in the smart card and computes user password. Sometime secrete key of the server stored in the smart card. If password is calculated with help of smart card, then attacker can act as a legal user.

**Countermeasure:**

User must keep secure their smart card and if any case smart card was stolen in that case designer must keep in mind of information in smart card so at least attacker could not able to extract information from smart card.

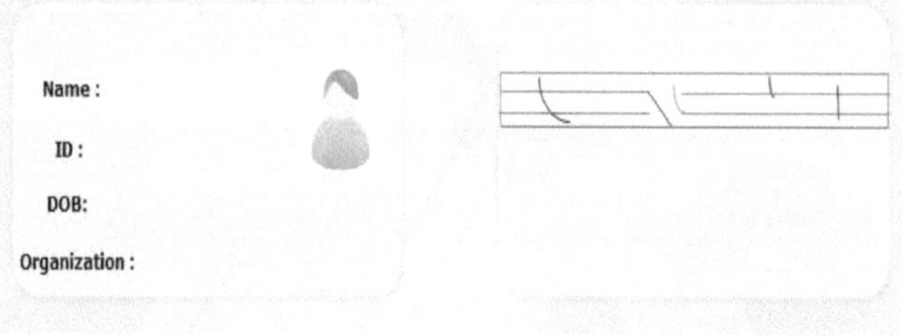


**Fig. 5.8 Stolen Smart Card Attack**

## 5.12 Session Key Agreement (A12)

After authenticating remote user and server to each other, communication between user and server will start and all messages will be in encrypted format, for serving this an session key is important. Both the user and server have same session key for particular transaction. So it is essential to have same session key between user and server is called Session key Agreement. If any scheme does not provide same session between user and the server then it does not provide session key agreement.

**Countermeasure:**

Design mechanism should provide proper session key agreement.

## 5.13 Server Spoofing Attack (A13)

When in Remote User Authentication scheme, if there is no facility of mutual authentication then there is a high probabilty of server spoofing attack. In mutual authentication scheme firstly remote user and remote server will authenticate to each other and then transmit the credentials. But in one-side authentication scheme, just server can authenticate the remote user but user can't authenticate the server so that any illegal server take part in communication channel between legitimate user and legitimate server and then capture secret credential of user and spoof the remote user.

**Countermeasure:**
Improve authentication scheme, use certificate for authentication, use encrypted message technique.

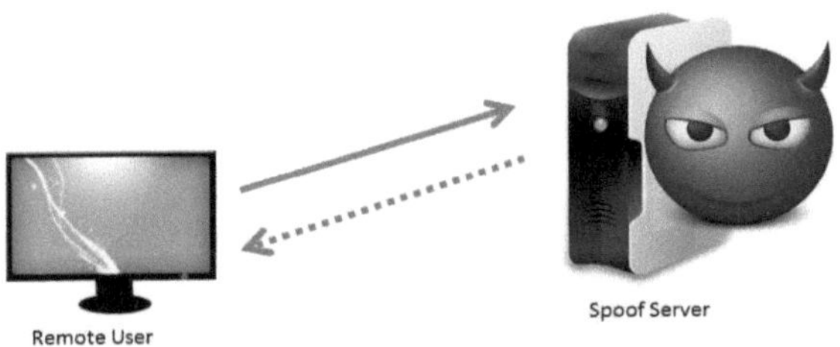


**Fig. 5.9 Server Spoofing Attack**

## 5.14 Denial of Service Attack (A14)

A DOS attack is an endeavor to make a user or server distracted for genuine clients and, at last, to bring the administration down. This is accomplished by flooding the server's call line with fake call. After this, server won't have the capacity to handle the solicitations of true blue clients. As a rule, there are two types of the DOS attack. The main structure is on that can crash a server. The another form of DOS attack only floods a service.

**Countermeasure:**
Use network layer mechanism to prevent this. And also use some kind of mechanism to identify each and every message transacted over network. Use intrusion detection system. Also use some kind of alarm when possible attack is performed.

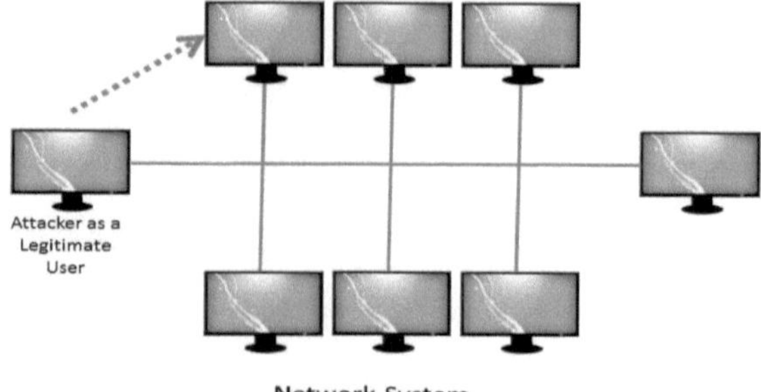


**Fig. 5.10 Denial of Service Attack**

## 5.15 Forward Secrecy (A15)

It is a feature of communication protocol, that provides guaranteeabout session keys that it will not be compromised, it means communication protocol have forward secrecy. Sometimes called Perfect Forward Secrecy or just FS. It also protect past session against future bargain of secret keys or password. Encrypted records in the past cannot be retrieved and decrypted should long-term passwords be bargain in the future by using Forward secrecy.

**Countermeasure:**

Use SSL combined with cookie management system. Over hubs prefer switches to avoid this [19].

## 5.16 Known Session Specific Temporary Information Attack (A16)

Leak of session secret values (Short-term secret) information should not compromise the new generated session key. The attacker can interpose the channel message and record it, which is transmitted over public channel. Then the attacker easily calculates the session key by using public parameters and lacked parameter.

**Countermeasure:**

Use SSL combined with cookie management system. Over hubs prefer switches to avoid this. Cryptography or secure protocol also helps to prevent. The remote connections and incoming connections can be minimized as to solve attacks [19].

# 6 Conclusion

Here, we have study some of remote user authentication systems. And survey on different aspects like types of possible attacks on remote user authentication system. We demonstrate the comparison of scheme regarding security parameter and computational complexity. This survey is used for reduce computational and communication cost of new designed scheme which is going to be publish. This paper will help to new researchers of remote user authentication framework to recognize shortcoming, security investigation and assaults of different frameworks. Thus, with the help of this paper one can propose their own systems with improvement of security against various attacks.

# 7  References


1. Leslie Lamport: Password authentication with insecure communication. Communication of the ACM, vol. 24, no. 11(1981) 770-772.
   DOI: 10.1145/358790.358797

2. T. C. Wu, Remote Login Authentication scheme based on a geometric approach, computer Comm. Vol. 18, No. 12(1995) 959-963.
   DOI: 10.1016/0140-3664(96)81595-7

3. M. -S. Hwang and L.-H. Li: A new remote user authentication scheme using smart cards. IEEE Transactions on Consumer Electronics, 46(1)(200) 28-30.
   DOI: 10.1109/30.826377

4. J.K. Lee, S.R. Ryu, K.Y. Yoo: Fingerprint-based remote user authentication scheme using smart cards. Electronics Letters 38 (12) (2002 June) 554-555.
   DOU: 10.1049/el:20020380

5. C.H. Lin, Y.Y. Lai: A flexible biometrics remote user authentication scheme. Computer Standard and Interfaces 27 (1) (2004) 19-23.
   DOI: 10.1016/j.csi.2004.03.003

6. C.-T. Li and M.-S. Hwang: An efficient biometric-based remote authentication scheme using smart cards. Journal of Network and Computer Applications, 33 (2010) 1-5.
   DOI: 10.1016/j.jnca.2009.08.001

7. X. Li, J.-W. Niu, J. Ma, W.-D. Wang, and C.-L. Liu: Cryptanalysis and improvement of a biometric-based remote authentication scheme using smart cards. Journal of Network and Computer Applications, 34 (2011) 73-79.
   DOI: 10.1016/j.jnca.2010.09.003

8. Ashok Kumar Das: Cryptanalysis and further improvement of a biometric-based remote user authentication scheme using smart cards. International Journal of Network Security & Its Applications (IJNSA), Vol.3, No.2, (March 2011).
   DOI: 10.5121/ijnsa.2011.3202

9. T.T. Truong, M.T. Tran, A.D. Duong: Improved ID-Based remote user authentication scheme using smart card. Wireless Communications and Mobile Computing Conference (IWCMC), (2013) 1672-1677.
   DOI: 10.1109/IWCMC.2013.6583807

10. A. K. Das: Analysis and Improvement on an efficient biometric-based remote user authentication scheme using smart cards. IET Information Security, vol. 5, no. 3 (2011) 541 – 552.
    DOI: 10.1049/iet-ifs.2010.0125

11. Y. An: Security analysis and enhancements of an effective biometric-based remote user authentication scheme using smart cards. Journal of Biomedicine and Biotechnology, vol. 2012, Article ID 519723 (2012) 1-6.
    DOI: http://dx.doi.org/10.1155/2012/519723

12. Muhammad Khurram Khan and Saru Kumari: An Improved Biometrics-Based Remote User Authentication Scheme with User Anonymity. BioMed Research International Volume (2013) 1-9.
    DOI: http://dx.doi.org/10.1155/2013/491289

13. Fengtong Wen, Willy Susilo and Guomin Yang,: Analysis and Improvement on a Biometric-Based Remote User Authentication Scheme Using Smart Card, Wireless Personal Communications, Ver. 80 (2015) 1747-1760.
    DOI: 10.1007/s11277-014-2111-6

14. Mrudula Sarvabhatla, M. Giri and Chandra Sekhar Vorugunti: A Secure Biometric-Based Remote User Authentication Scheme For Secure Data Exchange, Embedded Systems (ICES), (2014) 110-115.
    DOI: 10.1109/EmbeddedSys.2014.6953100

15. Li, X., Niu, J., Wang, Z., Chen, C.: Applying biometrics to design three-factor remote user authentication scheme with key agreement. Security and Communication Networks (2013)
    DOI: 10.1002/sec.767

16. Lee CC, Chang RX, Chen LA: Improvement of Li-Hwang's biometric-based authentication scheme using smart cards. Wseas Transaction on Communications, ISSN: 1109-2742, Issue 7, Volume 10, (July 2011).

DOI: http://dl.acm.org/citation.cfm?id=2064782

17. Younghwa An: Improved biometrics-based remote user authentication scheme with session key agreement. In: Kim, T.-H., Cho, H.-S., Gervasi, O., Yau, and S.S. (eds.) GDC/IESH/CGAG 2012. 351, vol. CCIS, Springer, Heidelberg (2012)307–315.
DOI: 10.1007/978-3-642-35600-1_46

18. Ankita Chaturvedi, Dheerendra Mishra, and Sourav Mukhopadhyay: Improved Biometric-Based Three-factor Remote User Authentication Scheme with Key Agreement Using Smart Card. Information Systems Security. Springer Berlin Heidelberg, (2013) 63-77.
DOI: 10.1007/978-3-642-45204-8_5

19. Jesudoss, A., and N. Subramaniam. "A SURVEY ON AUTHENTICATION ATTACKS AND COUNTERMEASURES IN A DISTRIBUTED ENVIRONMENT." IJCSE, vol 5.2 (2014).
DOI: http://www.ijcse.com/docs/INDJCSE14-05-02-061.pdf

20. Trupil Limbasiya and Nishant Doshi: A Survey on Attacks in Remote User Authentication Scheme. International Conference on Computational Intelligence and Computing Research, (Dec. 2014) 1-4.
DOI: 10.1109/ICCIC.2014.7238476

21. T. Hwang, Y. Chen, and C.S. Laih: Non-interactive password authentications without password tables, IEEE Region 10 Conference on Computer and Communication Systems, IEEE Computer Society, (1990) 429–431.
DOI: 10.1109/TENCON.1990.152647

22. Taher ElGamal: A PUBLIC KEY CRYPTOSYSTEM AND A SIGNATURE SCHEME BASED ON DISCRETE LOGARITHMS. Adv Cryptology-Lecture Notes Comput. Sci 196, 10–18 (1976).
DOI: 10.1007/3-540-39568-7_2

23. T. Hwang, Y. Chen, and C.S. Laih: Non-interactive password authentications without password tables, IEEE Region 10 Conference on Computer and Communication Systems, IEEE Computer Society, (1990). 429–431.
DOI: 10.1109/TENCON.1990.152647

24. M.S. Hwang, A remote password authentication scheme based on the digital signature method, International Journal of Computer Mathematics 70 (1999) 657-666.
DOI: 10.1080/00207169908804781

25. Ratha, N.K.; Karu, K.; Chen, S.; Jain, A.K.: A Real-Time Matching System for Large Fingerprint Database. Pattern Analysis and Machine Intelligence, IEEE Transactions on vol. 18, (1996) 799-813.
DOI: 10.1109/34.531800

26. P. Kocher, J. Jaffe, and B. Jun: Differential power analysis, Proceedings of Advances in Cryptology, (1999) 388–397.
DOI: 10.1007/3-540-48405-1_25

27. T. S. Messerges, E. A. Dabbish, and R. H. Sloan: Examining smart-card security under the threat of power analysis attacks, IEEE Transactions on Computers, vol. 51, no. 5, (2002) 541–552.
DOI:10.1109/TC.2002.1004593

28. "What is Biometric",2009.
Biometrics Research Group

# Acronyms and Glossary

| Acronyms | Meaning |
|---|---|
| AES | Advanced Encryption Standard. It is encryption decryption algorithm. |
| AIDE | Android Integrated Development Environment. It is android tool for computing JAVA code. |
| DOS | Denial of Services. It is one type of communication attack. |
| ID | Identity. Identity of user. |
| MD5 | Message Digest 5. It is hashing algorithm. |
| OTP | One Time Password. It is password that use only one time. |
| PIN | Personal Identification Number. It is secret value of user. |
| RSA | Rivest, Shamir and Adleman. It is public key cryptosystem. |
| RUA | Remote User Authentication. User can login from any remote location and can be authenticated by server. |
| SHA1 | Secure Hash Algorithm 1. It is hashing algorithm and it is highly secure compare to MD5. |
| SSL | Secure Socket Layer. It is cryptographic protocol used to provide security on computer network. |
| XOR | Exclusive or. It is a logical operation, where output is 1 if and only if different inputs are inserted. |

# About the authors

**Mr. Manish Shingala** received Bachelor of Engineering in Computer Engineering from Government Engineering College, Gandhinagar under Gujarat Technological University (GTU), Ahmedabad, India in 2014. He is currently pursing Master of Computer Engineering in Computer Engineering from Marwadi Education Foundation Group of Institution (MEFGI), Rajkot, India under GTU. He is interested in Research on Cryptography, Network Security, Smart card, smart grid, Three factor Authentication, CP-ABE. He is life member of Cryptology Research Society of India (CRSI), Kolkata, India.

**Prof. Chintan Patel** is a faculty in the Department of Computer Engineering at Marwadi Education Foundation, Rajkot since June - 2013, His main area of interest includes information security, Network security, Wireless Sensor Network and Internet of Things. In general he has completed M.Tech from SRM University , Chennai in 2013. He has worked as a Microsoft student partner for SRM University and published research paper on bio metric security. He is a IEEE Member and Life time member of Computer Society of India.

**Dr. Nishant Doshi** is a faculty in the Department of Computer Engineering at Marwadi Education Foundation, Rajkot since 2014. His main research interests includes algorithms, cryptography and remote user authentication, information protection in general. He has completed M.Tech from DA-IICT, Gandhinagar in 2009 and Ph.D. from NIT Surat in 2014. Along with active researcher, he is Editor-in-Chief of journals like IJCES, IJECEE, IJME, IJMES, and IJSCE. He is rewarded as *Young Scientist* from Venus International Foundation in year 2015.